Java:
2 Books in 1

*Tips and Tricks to
Programming Code
with Java*

*Best Practices to
Programming Code
with Java*

Charlie Masterson

actions solely under their purview. There are no scenarios in which the publisher or the original author of this work can be in any fashion deemed liable for any hardship or damages that may befall them after undertaking information described herein.

Additionally, the information in the following pages is intended only for informational purposes and should thus be thought of as universal. As befitting its nature, it is presented without assurance regarding its prolonged validity or interim quality. Trademarks that are mentioned are done without written consent and can in no way be considered an endorsement from the trademark holder.

Table of Contents

About this Bundle:

Congratulations on owning *Java: 2 Books in 1 - Tips and Tricks and Best Practices to Programming Code with Java* and thanks for doing so.

What you are about to read is a collection of two separate books on how to learn Java computer programming for the **intermediate level**.

Book 1:
Java: Tips and Tricks to Programming Code with Java

Here you will progress and learn intermediate level Tips and Tricks in learning about the Java programming language – to help you progress in your path towards Java mastery.

Book 2:
Java: Best Practices to Programming Code with Java

Here you will learn intermediate level best practices in order to write more efficient and better Java code.

Thanks again for owning this book!

Let us begin with the first book in the Java bundle:

Java:

*Tips and Tricks to
Programming Code
with Java*

Charlie Masterson

Introduction

Congratulations on owning *Java: Tips and Tricks to Programming Code with Java* and thank you for doing so.

The following chapters will discuss how you can use Java and what tips and tricks you are going to be able to use to further what knowledge you know on Java.

This book is mean to help you push past the most basic of knowledge and find easier ways for you to be able to use Java so that it can benefit you.

There are plenty of books on this subject on the market, thanks again for choosing this one! Every effort was made to ensure it is full of as much useful information as possible, so please enjoy!

Chapter 1: Hibernation in Implementations that are Standalone

Before you can begin to develop a project, you need to be able to have a strategy that is going to work well with the data that you are inputting into the program.

This is something that needs to happen no matter what project that you are working on so that you know where you are supposed to go instead of trying to do whatever and potentially mess up what you are trying to do. Not only that, but you are going to be able to use a framework that is going to be reliable and improve your project over the course of the time that it is used.

However, if you are using a framework that is unstructured and cannot be controlled you are going to end up having to continuously fix your project and spend money on trying to maintain it properly so that it does not change your code and mess up your entire project.

Basically, making sure that the framework that you decide to use is going to make sure that your project is protected and has the proper tools that are needed to debug your project.

The hibernate validator is going to a project that is open sourced and will help when it comes to demonstrating key features or composing rules that are going to be valid.

Hibernate Validator

The validator for the hibernate is going to be a foundation that is built solidly but will allow for you to write code that is flexible and lightweight for Java EE and SE. The validator is going to be able to help with different frameworks that are going to be used when working with Java but you are also going to be able to use frameworks that are going to standalone.

The Java SE standalones are going to become very important parts of the application on the server side for more complex heterogeneous.

To be able to use the validator so that you can build a component that is stand alone, you are going to want to make sure that you have JDK version six or higher installed on your computer.

Having this version is going to make it to where you can use the 5.0.3 version of the validator by downloading version five of the hibernate binary distribution package. Your directory is going to have all of the binaries that you need to make sure that you are able to build implementations that are stand alone.

The first example will show you how an ant built script is going to have the dependencies that are going to be for standalone implementations under the manifest section. This section is going to be required for the metadata that is outside of the code that you are writing. Because of the ant build, you are going to have all of the validator's that are dependent JARs through the header of class-path for the manifest file.

Example: manifest section with ant build and dependencies

```
CODE:

<manifest>

<symbol title = "I made this" amount =
"${ name. user} " / >

<symbol title = primary-class" amount =-
"come.utility.validationutilities" / >

<symbole title = "class-path" amount =
"library/hibernate-validator-dic-5.0.3.
last.bottle libarary/ hibernate-
validator-mark-processor- 5.0.3. last
bottle library/validation-cdi 1.1.0 last
jar library/ classperson – 1.0.0 glass
library/ javae dl – 2.2.4 glass library/
javae.le-dpi- 2.2.4 glass library jhead-
listing- 3.1.1. DP. Glass library/
system4k- 1.2.17 glass . " / >

< / manifest >

<metainj directory - $ {workplace_ deal_
route} / meta idj" additionally = *.mlx"
/ >
```

Annotation declaratives and definitions for constraints

With version five of the hibernate validator, you are going to have an implementation that is open for the JSR bean validation version 1.1. Annotations for the declaratives and the definition for constraints is going to have two highlights for the bean validation framework that has been updated.

When you look at the rules for validation that are made through the syntax for declaratives is going to improve how your code is read.

Example: declarative annotations

```
CODE:

Open class place {

@codeishere

Closed string nameofwholiveshere ;

...

@morecode

@pattern ( regapd  " [ D-Ej-d] \\ j[ D-
Dj-d] \\e? \\ j]D-Ej-d\\\ j" , readme =
"this is not the correct address for this
person." )

Closed string codeforcity

@codehere

Closed string city;

@morecode

@pattern (regapd = FH|ME|EL| EJ| AL| EE|
ME| DI| AO| EM | NE| OI| LA}" , readme =
"this is not the correct city for this
zip code" )

Closed string zipcode ;

...
```

The validation method is going to have conditions that are going to be defined based on conditions that were set before so that the code is improved on readability. The rules for this are going to be easier for you to understand being that you will only need to take notice of the annotations and methods that are being used.

Declarative styles are going to be delete the necessary need to have a model for execution due to transitions of the different states all while analyzing the conditions that are going to come before and after the code.

Constraints will apply to the objects that fall under the @valid annotations.

Chapter 2: SharePoint Services on the Internet that Use a Java Client

For this chapter you are going to learn CRUD operations for SharePoint documents that use a Java client. It is going to focus on the web services that are going to be found in the copy and list services that you can use with Microsoft.

CRUD services can also be used with CAML which stands for Collaborative Application Markup Language. It is based on XML that is going to utilize the different methods that are going to be used with copy and lists.

On top of that you are going to be able to construct a CAML structure that is going to be able to pass through the different conditions that are going to be set for the properties that are assigned to the objects in the code.

Background

In the event that you are working with a large amount of data that may be in different forms, you are going to be able to look at that data in a compounded file so that you can make any decisions that you need to make. There are different kinds of monitoring that is going to happen in this kind of environment.

An audit is going to make sure that the data you are working with is going to be in a constant state and then these reports can be sent out to the people that need to have them in order for them to be saved to SharePoint.

But, the engines that you are going to use are going to need to have an output writer concept that is going to help set up writing in SharePoint. This helps with the writing that happens in databases and the servers that are SMTP so that you can avoid having to do everything manually.

Communicating with SharePoint

When you are working with SharePoint you are going to be using code that you write on your own for the solution that you need to reach. The code is going to be saved onto the Microsoft SharePoint website which is also code generated.

There are going to be two services that you are using with Microsoft, because of this, you are going to be accessing different services that are offered through SharePoint. They will be communicating through two different locations.

Location one will be:
https://server/site/_vti_bin/Lists.asmx
The second location:
https://server/site/_vti_bin/Copy.asmx

Generating packages on web services
In order to generate the web service packages that you are going to be using, you are going to be using a wsimport function that you can find in the bin directory when you look at how Java has been installed. That is going to be based on

the fact of you have installed version 1.6 of Java.

SharePoint is going to run on HTTPS which may end up causing you to find a few problems with wsimport when you look at the server that you are using thanks to the two different URLs that you are going to use.

In that case, you are going to end up getting an error that will look similar to this.

CODE:

```
[ERROR] bright.
Protected.validator.validatorException:
XPIK follow building untrue.
Bright.procted.offered.certified.BrightCe
rtFollow

BuiltException: cannot find certification
that is valid enough to get to the
target.
```

This error happens because the file that you are using does not have the proper certification that it needs for the site to run. How you are going to get around this is to download a file locally for WSDL.

Example: Copy service code

CODE:

```
C : \ parma > "% JAVA_NATIONAL% \ nib \
swimport " -e . -a come.microsoft.
schemas. Sharpoint. Clean - hold -addmore
-Anything Copy. Swld

Parsing DLSW
```

```
[ALERT] CLEAN opening "copyclean91" :
please us a different holding than this

Section 34 of file : / C: / parma/ Copy/
swdl

Code is being created
```

Example two: lists service code

```
CODE:

C : \ parma > "% JAVA_NATIONAL% \ nib \
swimport " -e . -a come.microsoft.
schemas. Sharpoint. Clean - hold -addmore
-Anothing list. Swld

Parsing DLSW

[ALERT] CLEAN opening "listclean91" :
please us a different holding than this

Section 34 of file : / C: / parma/ list/
swdl

Code is being created
```

After the code, has been created, then it can be
added to the solution that you made and use it.
The compile option can be changed to a
different command. However, doing this is
going to make the class files be created
differently inside of the source that you are
using while allowing Eclipse to compile them
only if we have written the source code out
already.

Security

So, that SharePoint does what you want it to do, you are going to need to change your process so that you can add in Axis2. Axis2 is going to have a few issues, but you can overcome the issue by using the JCIFS along with Axis2. However, you may think that this is going overboard, but it is going to do what you want it to do all while making it easier to do what you want done.

Should the website that you are using have the cacerts file and have it updated to where it is going to be, you can use this as long as the certification for the site is up to date as well.

CODE:

```
Ajavae.web.lss.trustbought - you will use
this to update the files that you are
using.
```

Chapter 3:
SOA Integration Using Apache Camel

The Apache Camel is an integration service that is geared towards the architecture of the program that you are running. Camel lets you configure the integrations so that it works for web services such as message transformations, routing, and the handling of exceptions. All without having to use too much code from Java.

The integration in Java is going to mainly be used for the priority of projects that are looking to connect the end points for multiple web services quickly so that they work efficiently and are easy to maintain.

When you look at the development of the standpoints, the integration is going to be a challenge that you have to be willing to work for. However, you can make it easier by using the Apache Camel framework.

The API and different components for the Apache Camel are going to be mostly recognized when it comes to enterprise integration patterns or EIPs for short. The EIPs make it easy to do the tasks that are required when it comes to working with integration for the connection of web services like XSL, logging audits and much more.

Enterprise integration example

Think about working for a major airline that is trying to make it easier for their customers to be able to make reservations. One of the simplest ways to do this now is online because everything is online now days. When creating the service, it is going to compare multiple airlines and their prices so that the customer can get the best price for when they want to travel.

There are going to be several web services that are going to be included in this because the airline that you are working for is going to have to pull the prices from their competing airlines websites so that they can give it to their customers.

You are going to be making a list of quotes based on what the customer is looking for in their airline experience. The portal that the customer is using is going to be specific to them and give them everything that they are wanting from their destination to the best price that they can get for that fare and when they are going to reach their destination.

Writing out the custom Java code is going to require a lot of Java code so that you can map everything to process the request from the user properly.

It is best that you are able to map out the request from start to finish by using threads in Java to set up a parallel web service that is going to get all the services involved in supporting the task.

Apache Camel is going to use a framework that will allow shortcuts to make the programming easier. All you are going to need to do is leverage the API components for Camel and therefore design a web service integration task that is going to work with the EIPs. Instead of using the code for Java, you are going to use DSL or Domain Specific Language so that it only works with the domain that it is set up to work with.

Apache Camel is going to make DSL code based on Java so that it can route and configure the integration layer the way that it is supposed to be.

Continuing with our airline example, you are going to see through the portal that your customer is going to send a request through the starting location and the airline that they would prefer to use. From there, the code that you have written into the website is going to send back the airlines and destination information that they are looking for so that they can compare prices and airlines.

If they have decided on what airline they want to use, then they are going to use the reservation service to accept the quote that came through to them based on the preferences that they inputted.

Java code is going to need to be written for the processors so that the routes that the customer takes are going to not be compromised by the website.

Terminology

A route in the Apache Camel is going to be a chain of processors that a message is sent through between two points. The preferred route is going to go between the endpoints. The messages are going to be kept in a component that is going to be referred to as the exchange.

The runtime that it takes for Camel to get the message from point at to point be is going to be the Camel Context.

Components for Camel SOA

CXF Endpoint: Apache CXF is going to be used for web services and their implementations. The CXF component is going to be used when configuring the web services that are being used and help to determine which route they are going to take.

Camel Bean: the component that uses JavaBeans. The bean is going to be used when setting up custom processes that have steps that need to be followed.

Camel Direct: two routes are going to be connected through a synchronous invocation.

Camel XSLT: the XSLT sheet will be loaded by the springs resource loading technique. This component is going to be used to handle the request and response transformations from web servers.

Camel Multicast EIP: the EIP pattern is going to send messages to different places that need to be processed. From here you can use multicasting to send the requests.

Camel Aggregator EIP: the pattern is going to combine two different results that have messages that are similar into one message that will be sent to the user. The aggregator is going to combine the responses so that the client is not getting overwhelmed.

Camel WireTap EIP: with wiretap, you can inspect the messages while they go through the system to the user. The EIP is going to log the messages with the points that are appropriate for it.

Web interfaces

WSDL files are going to be tied to the schemas for XML in web service interfaces when you look at the source code.

There are eight steps that you are going to use to implement the SOA layers.

Step one: Generate Java types for WSDLs

You are going to use the cxf codegen plugin that works with Maven along with the wsdl2java took that is going to assist in generating the Java code that you need for the different types of WSDLs that you are going to find in the source code.

In order to execute this code you are going to use the cmd: mvn generate sources code.

From here CSF is going to generate the code that is needed to place the artifact into the proper directory that it belongs in. From there, Maven can be used to build the artifacts that you are going to use.

Example

```
< plugin >

<id for group> gor. Camel. Cxf < / id for group >

<id for the object> cxf codegen plugin < / id for the object>

<version> 2.8.0 < / version>

<carry out the code>

<number> create source < / number>

<series> create source < / series>

<configure>

<origin> $ { bottomry } / cor/ primary / java < /origin>

<Wsdlchange>

// reservation made by user

<wsdl> $ {bottom ry}/ cor/ primary/ resource/ wsdl/ reservation </wsdl>
```

```
</wsdlchange>
<wsdlchange>

//reservation made on airline D

<wsdl> $ { bottomry } / cor / primary/
resource/ wsdl/ reservation on airline d.
wsdl </ wsdl>

</wsdlchange>

<wsdlchange>

// reservation by airline F

<wsdl> $ {bottomry}/ cor/ primary/
resources/ wsdl / reservation on airline
f wsdl </wsdl>

</wsdlchange>

</wsdlchanges>

</ configure>

<achieve>

<achieve wsdl2java </achieve>

</achieve>

</ carry out>

<plugin>
```

Step two: Configure Camel CXF end points

Here is where you are going to need to define
the end points for your code. The end points
are going to be used in the routes that you use
for Camel.

You need to remember though that the interface that you are using is going to need to be generated by the end point that you have already put in place.

Camel cxf end point will be the default end point in your code so that you can get XML in its raw form. You will need to enable the validation so that you can validate the requests that come in through XSD in the WSDL so that you can reduce time that needs to be invested into the validation.

Example

CODE:

```
// end point for the reservation made on
airline D
<cxf: endpoint number = endpointairlineD

URL = "http:// nationalhost: 9898/
reservation made" servicegenre = "mor.
Small. Rira. Recall. airlineDquote"

Wsdladdress = "wdsl/ reservation for
airline d wsdl">

<cxf: ability>

<enter key = informationinput" number =
"pay" / >

</ cxf: ability>

</ cxf: cxfend>

// end point for the reservation made on
airline F
<cxf: endpoint number = endpointairlineF
```

```
URL = "http:// nationalhost: 9898/
reservation made" servicegenre = "mor.
Small. Rira. Recall. airlineFquote"

Wsdladdress = "wdsl/ reservation for
airline f wsdl">

<cxf: ability>

<enter key = informationinput" number =
"pay" / >

</ cxf: ability>

</ cxf: cxfend>
```

As for the explanation of the rest of the steps:

Step three is so you can make the proper routes with the use of multicast.

Step four lets you see the messages as they go through the system.

Step five is where you can write your messages by using XSLT.

Step six uses wiretap.

Step seven will create the context for XML.

Finally, step eight will execute the code you have written.

Chapter 4:
Orthogonality

Using orthogonality is going to make it easier to maintain your program's software therefore making it easier to understand for your users.

Orthogonality comes from the Greek word orthogonios which literally means "right angled". This is normally a word that is used when you are talking about the difference between a set of dimensions.

As you look at a graph, you are going to notice that an object is going to move along the x axis but only if it is occupying a space of three dimensions.

The coordinates are not going to change. But, if you change the dimension of one of the coordinates, you are going to need to change the other dimension because there cannot be side effects on other dimensions just because you changed one.

That is why orthogonality is used for describing software designs that are modular and maintainable as you think about the system in a multi-dimensional space.

Orthogonality is going to be what assists a developer to make sure that the changes are not going to have side effects on other parts of the system.

Dimensions of Log4j

When you use logging, you are doing nothing more than using a fancy version of the println()

function that you use in Java, but Log4j is going to have a package that is going to change the mechanics of how the Java platform is going to respond.

Not only that: it also allows developers to do different things such as:

- Creating new layouts so that events for logging are defined by a string.
- Appenders are going to be logged differently. This is not limited to consoles but also to locations of networks, logs for operating systems, and more.
- Control will be centrally located depending on how much information has to be logged.
- There are now layers for things like debug and errors.

Log4j includes other features that you can explore as you work with Java.

Log4j types considered like aspects

The appender has levels and there are three aspects of it that you are going to see on the independent dimensions of the program.

- Level: the events that are logged by the program
- Appender: the data that is displayed or stored

- Layout: how the data is presented to the user.

You have to look at the aspects all together instead of individually and they have to be on a three-dimensional space. Every point for the aspect is going to be part of the space that is going to configure the system the way that it is supposed to be.

Example:

```
CODE:

// logging started!

Logging logger = logging . the logging
aspect ("No") ;

Appender appended = old the appender for
the system() ;

Layout display = old gor camel log4j
CTCTdisplay()

Appender. Howdisplayed (displayed) ;

Logger. Showappender (appender) ;

Logger. Howhigh (information on how high
it is) ;

// logging started!

Logger. Alert (" You shall not pass!") ;
```

The code in the example is an orthogonal code and it is going to enable you to change the layout, appender, and level of the aspect without ever causing your code to break therefore it continues to be functional.

The design for orthogonal code is going to have points that are going to have enough space on the script that is going to make it valid so that you can configure the system.

The orthogonality is a concept that is powerful and is going to establish a mental model that is going to be somewhat simple to understand even though you are using it for applications that are more complex.

If you want a situation that is going to be useful to use orthogonality, then you are going to use it when you are testing the code's functionality and levels so that you can fix the code that is not going to work with the appender or the layout.

The use of orthogonality is going to make sure that there are no surprises with your code and that they all work the same on each level. This is going to be true when you are using log4j and other tools and utilities that are designed by third parties.

Thanks to the complexity of the orthogonality, program's software is going to be somewhat like the dimensions that you use when you are working with geometry.

Code and designing it for orthogonality

Now that you know what orthogonality is, you are going to need to know how to design and code it so that you can use it in your programming. The whole point behind it is that you need to use abstract concepts.

Every dimension that you see in an orthogonal system is going to have one aspect for each program. Some dimensions are going to have types that are going to be one of the most common solutions for using abstract types.

Every type will have a dimension that is going to represent the points that are given inside of the dimension. Abstract types are not going to be directly tied into instantiated or concrete classes.

There are some classes that you do not need to use. Like you do not need to use concrete classes whenever the type you are using is just a markup and not showing any encapsulated behavior.

However, you can just look at the type to see if it represents the dimension itself and if it is predefined for a set of instances that are fixed in place. It will even work for variables that are static.

A general rule of thumb to remember when you are working with orthogonality is to try and avoid the references that are going to be for concrete types that will represent a different aspect for the program that you are using.

Keeping to this rule is going to allow you to write code that is more generic code that is going to work in all instances that you are going to be putting it into.

The code can have references to properties just as long as they are part of the interface that is used when defining the dimension.

For example, if you look at the aspect type of layout that defines a different method, then this method is going to give you a Boolean that will indicate if the layout has the ability to be rendered with stack traces or not.

At the point that an appender uses the layout, it is going to be fine to write code that is conditional for the method that you are using. That way that the appender for the file will print the exception of stack traces that are going to use a layout that cannot normally handle exceptions.

The layout implementation is going to refer to when you are using a level in particular that is going to be used when you are logging events.

For example, if you want to log a level error in a layout that is using HTML, then you can wrap the log message in tags. But, the error is going to define the level you are working with by representing the dimension.

References can still be avoided based on implementation classes that you use with other dimensions. So, if you use a layout then you are not going to need to know what sort of layout you are using.

Violating orthogonality

Log4j is one of the best examples that you are going to have for orthogonality. But, some parts of log4j are going to violate the rules in orthogonality.

There is an appender that is going to log the relational database. Since there is a scale and popularity issue with the database, it makes it easier to search for the logs with SQL.

There is an appender that is based on fixing the problem that you will run into when it comes to logging a database that is rational by turning the events that are logged into the insert statements with SQL. Therefore, the problem with the pattern layout is solved.

The patternlayout is going to have a template that is going to give the user flexibility that is going to configure the strings to be created each time that an event is logged. The template will define the string and all the variables that are included in it.

Example:

```
CODE:

Pattern for the string =
```

```
"%d [ @ %a   {ss:mm:HH yyy MM dd} no %l]
%l%j" ;

Layout display =

Old gor. Camel. Log4j layout for the
pattern (pattern) ;

Appender. Layout setup (display) ;
```

The appender will use the layout with the pattern that is going to define the statement.

Example:

```
CODE:

National untrue holdSql (string d) {

Statement for sql = d ;

if (display() == false) {

This display (old layout for the pattern
(d)) ;

}

else {

(( layout for the pattern) pattern())
conversion pattern (d) ;

}

}
```

There is an implicit assumption that is built into the code for the set that you are using in this example that will define the appender. When looking at the orthogonality code, this is going to point to a lot of different aspects of a three-dimensional cube that is going to show the layout with the patternlayout that is not going to be represented in a system configuration that is valid.

Any attempts that are put into place for the SQL are going to have to use a layout that is different so that there is a different exception being used for the class cast.

If that is not enough reason for you, then another reason is that the appender's design is quite questionable. When you use the layout for the pattern, the template will be bypassed.

But this is not always a good thing because the JDBC will compile all of the statements that were prepared previously and this is going to end up leading to improvements in the performance that are not going to be easy to fix. The appropriate approach for this though is to control what your layout does by overriding it.

Example:

CODE:

```
National open display set (layout
display) {

if ( display instanceof layout pattern) {

Superb layout setup (display) ;

}
```

```
else {

Set a new argument that is not valid ("
layout is not going to work") ;

}

}
```

Sadly, you are going to find problems with this example as well. There is a runtime exception that is going to make the application using this method unprepared to catch the mistake. Essentially, you are going to use the layout method, but it is not going to have a runtime exception that is going to work with it and the guarantee is going to be weakened.

If you look at the preconditions that are set up, then the layout is going to need stronger preconditions so that the method can override it. If you do not do it that way, then you are going to violate the object-oriented core that is going to be meant just for the principle.

Workaround

Since there is no really easy solution for fixing the appender, you are going to end up finding that there are deeper problems that you have to deal with. Because of this, the level of abstraction that you are going to use is going to happen when you are designing the abstract types for the core which are ultimately going to need to be fine-tuned.

The core method is defined by the layout and how it is formatted. The method will end up returning a string to you which is going to have a database that is based on tuples and not strings.

A solution that you can use is one that is going to be based on a data structure that is more sophisticated for the return format. You will need to imply the overhead in any situations that you are going to want a string to be returned to you. Also, the objects that are intermediate are going to be used when created before dealing with the garbage collected performance that is working on the framework for logging.

When you have more than one return type that is sophisticated, you are also going to make log4j more difficult for you to understand. The ultimate goal is to make everything as easy to understand as possible.

Yet another solution is to use abstracts that are layered and involves using two abstract types that will end up extending the appender. The only appender that you should be using is the JDBC appender because most other appenders are going to have implementations that have to be implemented so that they work the way that they need to.

The biggest drawback that you are going to see with this one is that the complexity is going to increase yet again. Not only that, but the developers of the appender will need to make sure that they have all the information before they make the decisions that are needed to be

made for the level of abstraction that they are
going to use.

Chapter 5: JavaFX Applications That Can Be Used in Multiple Environments

When you are using version 2.0.2 of JavaFX along with SDK you are going to be allowed to deploy applications inside of multiple environments along with applications that are going to stand alone thanks to Java web start or a web page that is embedded.

Preparing JPadFX for deployment

There are source files that you are going to have to use with the JPadFX application.

After all of the files have been placed into the directory that you are using, then you are going to be allowed to execute the commands that you put into the application.

Example:

CODE:

```
Javae - pc "C: \ project source \ oracle
\ javafx 2.0

DKS\ tr\ library\ trxfj. Glass" ; . -q in
jpadfx. Java
```

Please note that this is only going to work if you have version 2.0.2 installed on your computer.

Deploying JPadFx as standalone applications

With the version that you should have downloaded on your computer, you are going to have a tool that is going to be for inserting your commands so that the application is deployed almost automatically. When you look at it, there is a completely different directory that is used for this version of the program.

After the source files, have been compiled, then you can use the javafxpackager so that you can deal with the class files along with the launcher into a new file.

Example:

```
Javafxpackager makeglass applicationtype
jpadfx directory crs in directory outside
in inside file jpadfx. Glass. D
```

Terminology

- makeglass are going to tell the program to create a new file in that directory.
- Applicationtype: the main class is going to be identified with the launcher
- Dircrs: the directory will be identified.
- In directory: the name of the file will be identified
- -d is going to tell you what the output is going to be and where the error message is going to be displayed.

Should the commands be successful then you are going to be introduced to a new function. The simplest way to deal with this is to use the jar tool. But, that is only going to work if your current directory has the proper tools and the subdirectory tied to it so that the command can be carried out and stored in the directory.

Example:

```
CODE:

Glass fu

In \ jpadfx. Glass icon.png
```

It is at this point that the stand alone will be run with Jpadfx so that you can test the file's veracity. But you are going to need to switch out of the directory and into a new directory that has the correct command that needs to be executed.

Example:

```
CODE:

Java -pc "C: \ project source\ oracle\
javafx 2.0

DKS\ tr \ library \ jfxrt. Glass" ; . -
glass jpadfx.glass
```

This command will add the class paths to the correct files so that the runtime is loaded correctly.

Deploying JPadFX in a web page

The javafxpackager can be used in creating HTML pages along with deploying them with the correct files so that scenarios can be handled when it comes to embedding the applications in a web page or running it with Java web start.

Example

```
Javafxpackager -execute -in directory in
-in file jpadfx -how wide 500 -how tall
500 - directory crs in crs source
jpadfx.glass

-class of application jpadfx -title "jpad
fx" -name "jpadfx is equal to jpad" -
manager "Timothy Mason" - d
```

Terminologies explained:

- **Execute:** the packager will be executed with HTML and JNLP files
- **Directory in**: identifies which directory the file is stored in
- **Infile**: the file name will be identified
- **How wide**: the width of the page
- **How high**: how tall the page is
- **Directory crs**: where the file is coming from
- **Crs source**: the file is identified

Chapter 6:
Controlling the
Flow of Java

Any programming that you have done before this point is going to be known as sequential programming which basically means that the code you have entered was read from top to bottom, every line being read the way that it is written. However, not all programs are going to work like that.

Some code has to be run but only in the event that a condition has been met. For code, you may want your client to be of a certain age.

This is going to be where you will control the flow of how the program operates through conditional logic.

Conditional statements are going to use the word if mostly, but there are other words that can be used to create conditional statements.

If statement

The code is going to be executed whenever something happens instead of waiting for something else, which is going to be most common when it comes to programming, therefore the if statement was created. An if statement is going to look like this:

```
CODE:

if ( statement ) {
}
```

The word if is going to be lowercased before your condition is paced in a set of brackets. The curly brackets are going to be used to divide the chunks of code.

This code is going to be executed only in the case that the condition has been met.

Example:

CODE:

```
if( client < 18) {
}
```

This condition states that if the client is less than 18. Instead of typing it all out, shorthand is going to be used so that the chunk of code is shorter and easier to read.

Example:

CODE:

```
if( client < 18) {
      // display message
}
```

Therefore, should the client be over 18, the code set in the two sets of curly brackets is going to be jumped over while Java proceeds how it is supposed to. Anything that is in the middle of the curly brackets will be carried out only if the condition has been met.

Another short hand notion is going to be to point the triangle to the right which means greater than.

Example:

```
CODE:

if( client > 18) {
      //display message
}
```

The code is the exact same except that the condition is going to apply to those who are over 18.

If ... else

Instead of an if statement being used, an if then statement can be used.

```
if( condition_to_test) {
}
else {
}
```

The first line will still start with if while being followed by the condition. The curly brackets are going to divide different parts of the code.

The second choice is going to go after the else with its own curly brackets.

There are only two choices that you are going to have, either your client is going to be younger than 18 or older than 18. Your code will have to match what you want to happen.

If ... else if

More than two conditions can be tested at once. If you want to test a certain range of ages, you are going to use an if else if statement.

The syntax is going to be:

```
If(condition_one) {
}
Else if ( condition_ two ) {
}
Else {
}
```

The newest part of the code is going to be the else if (condition_two) {
}

When the if statement is sent, the next is going to be followed by round brackets. The second condition is going to be in between two new brackets but it is not going to be caught by the first condition.

The code is still going to be sectioned off with curly brackets with every if else if statement that you have each having its own curly brackets. If you miss one then you are going to get an error message.

The conditional operators that you need to use are:

Greater than (>)
Less than (<)
Greater than or equal to (>=)
Less than or equal to (<=)
AND (&&)
OR (||)

A value of (==)
NOT (!)

The two ampersand symbols are going to be used to test for more than one condition at the same time.

Example:

CODE:

```
else if ( client > 18 && client < 40)
```

With this code, you are going to want to have the knowledge need to know the age of the client and to see if they are older than 18 but younger than 40.

You are only trying to see what is inside of the variable that the client inputs. The client has to be older than 18 but younger than 40. Your client has to meet both conditions since the AND operator was used.

Nested if

If statements can be nested just like if else and if else if statements. An if statement that has been nested is going to be putting one if statement with another.

Example:

CODE:

```
if( client < 19) {
```

```
        system.out.println ( " 18 or
younger") ;
}
```

To check and see if your client is over sixteen there has to be another if statement placed inside of the if statement that you have already created.

The first if statement will be where the client is going to be caught if they are less than 19 and the second one will narrow it down so that they are over 16. When you want to print a different message, you can use an if else statement.

Example:

CODE:

```
if( client < 20) {
if( client > 20 && client < 23 ) {
System.out.println( "You are 21 or 22") ;
}
else {
System.out.println (" 22 or younger") ;
}
}
```

The curly brackets that are placed in the code have to be exact or the program is not going to be run. Nested if statement are tricky to learn, but they can be narrowed down.

Boolean values

Boolean values are only going to be true or false, 1 or 0 or even yes or no. Java has a variable that is done for Boolean values:

Boolean client = true ;

You are not going to need to type int, string, or double you are going to need to type Boolean by using a lower case b. the name of the variable will come after it has been assigned to the value of true or false.

The assignment is going to have one equals sign. However, should you evaluate the variable and it has a value of something then you are going to need to use two equals signs.

Example :

```
CODE:

Boolean client = true;
if ( client == true) {
System.out.println ("it's true") ;
}
else {
System.out.println("it's false") ;
```

The if statement is going to see if the variable entered by the client is true. The else will check to see if it is false.

You are not going to need to say "else if (client == false)". When something turns out to not be true, it is going to obviously be false.

You are going to use else since there are only going to be two choices when it comes to Boolean values.

The other operator that is used for conditions is the NOT operator, this is used with Boolean values.

Example:

```
Boolean client = true;

if( !client) {
System.out.println ("it's false") ;
}
else {
System.out.println ("it's true") ;
}
```

This is the same as any other code that is used with a Boolean except for the statement "if (!client) {".

The NOT operator is going to come before the client's variable. This operator will use a single exclamation mark but it will fall before the variable that has to be tested.

The code is going to be tested for negation which means that it is not what it actually is.

Since the client variable has been set to true, the !client will end up testing for false values.

If the condition is set to false, then the NOT operator will test for values that are true.

Conclusion

Thank you again for owning this book!

I hope this book was able to help you to gain more knowledge on the Java programming language and use it to apply to your programming requirements.

The next step is to continue programming as you learn more topics for your Java programming skillset.

Finally, if this book has given you value and helped you in any way, then I'd like to ask you for a favor, if you would be kind enough to leave a review for this book on Amazon? It'd be greatly appreciated!

Thank you and good luck!

Java:

*Best Practices
to Programming Code
with Java*

Charlie Masterson

Introduction

I want to thank you and congratulate you for reading my book, *"Java: Best Practices to Programming Code with Java"*.

This book contains proven steps and strategies on how to write Java code that doesn't kick up errors, is neat and tidy and has a high level of readability. If your code looks familiar, it will be far easier to understand and even easier to cast your eye over to see if there are any obvious errors. To make sure your code looks familiar, there are a set of best practice guidelines that you should follow; not only will this help you, it will help other developers who look over your code as well.

This book assumes that you already have a level of familiarity and experience with Java computer programming language. My aim is to help you further your knowledge and have you programming like a pro in no time at all. If you are a completely new to the language, please familiarize yourself with the basics before you run through my best practice guide.

Thanks again for purchasing this book, I hope you enjoy it!

Chapter 1: Formatting Your Code

Really and truthfully, the first place to start, before we get into the nitty-gritty, is in how to format your Java code. There are several "rules" that you should follow to make sure that your code is readable and clean:

Indentation

All Java code indenting uses spaces rather than tabs and each indent is 4 spaces. The reason for this is because, like all other computer programming languages, Java works well with spaces. Some programs mix up spaces and tabs, leaving some lines indented with tabs and some with spaces. If you set your tabs to 4 and your file is shared with a person that has theirs set to 8 it is all going to look a mess.

When you use matching braces, they must vertically line up under their construct, in exactly the same column. For example:

```
void foo()
{
    while (bar > 0)
    {
        System.out.println();
        bar--;
    }

    if (Cheese == tasty)
    {
        System.out.println("Cheese       is
good and it is good for you");
    }
```

```
    else if (Cheese == yuck)
    {
        System.out.println("Cheese  tastes
like rubber");
    }
    else
    {
        System.out.println("please    tell
me, what is this cheesel'");
    }

    switch (yuckyFactor)
    {
        case 1:
            System.out.println("This    is
yucky");
            break;
        case 2:
            System.out.println("This    is
really yucky");
            break;
        case 3:
            System.out.println("This    is
seriously yucky");
            break;
        default:

System.out.println("whatever");
            break;
    }
}
```

Use Braces for All `if`, `for` and `while` Statements

This is the case even if they are controlling just a single statement. Why? Because when you keep your formatting consistent, your code becomes much easier to read. Not only that, if you need to add or take away lines of code, there isn't so much editing to do. Have a look at this example – the first three show you the wrong way while the fourth shows the right way to do it:

Bad Examples:
```
if       (superHero       ==       theTock)
System.out.println("Fork!");

if (superHero == theTock)
    System.out.println("Fork!");

 if (superHero == theTock) {
     System.out.println("Fork!");
    }
```

Good Example:
```
    if (superHero == thetock)
    {
        System.out.println("Fork!");
    }
```

Spacing
Whenever you have a method name, it should be followed immediately with a left parenthesis:

Bad Example:
```
foo (i, j);
```

Good Example:
```
foo(i, j);
```

A left square bracket must follow immediately after any array dereference:

Bad Example:
```
args [0];
```

Good Example:
```
args[0];
```

Make sure there is a space on either side of any binary operator:

Bad Examples:
```
a=b+c;
 a = b+c;
 a=b + c;
```

Good Example:
```
 a = b + c;
```

Bad Example:
```
 z = 2*x + 3*y;
```

Good Examples:
```
 z = 2 * x + 3 * y;          !
 z = (2 * x) + (3 * y);!
```

Whenever you use a unary operator it should be followed immediately by its operand:

Bad Example:
```
 count ++;
```

Good Example:
```
 count++;
```

Bad Example:
```
i --;
```

Good Example:
```
i--;
```

Whenever you use a semicolon or a comma, follow it immediately with a whitespace:

Bad Example:
```
for (int i = 0;i < 10;i++)
```

Good Example:
```
for (int i = 0; i < 10; i++)
```

Bad Example:
```
getOmelets(EggsQuantity,butterQuantity);
```

Good Example:
```
getOmelets(EggsQuantity,
butterQuantity);
```

Write all casts without spaces

Bad Examples:
```
(MyClass) v.get(3);
( MyClass )v.get(3);
```

Good Example:
```
(MyClass)v.get(3);
```

When you use the keywords `while`, `if`, `for`, `catch` and `switch`, always follow them with a space:

Bad Example:
```
if(hungry)
```

Good Example:
```
if (hungry)
```

Bad Example:
```
while(omelets < 7)
```

Good Example:
```
while (omelets < 7)
```

Bad Example:
```
for(int i = 0; i < 10; i++)
```

Good Example:
```
for (int i = 0; i < 10; i++)
```

Bad Example:
```
catch(TooManyOmeletsException e)
```

Good Example:
```
catch (TooManyOmeletsException e)
```

Ordering of Class Members

When you use class members, they must always be ordered in the following way, without exception:

```
class Order
{
```

```
// fields (attributes)

// constructors

// methods
}
```

Line Length

Try not to make any lines more than 120 characters long. This is because most text editors can handle this line length easily; any more than this and things start to get a little messy and frustrating. If you find that your code is being indented way off to the right, think about breaking it done into a few more methods.

Parentheses

When you use parentheses in expressions, don't just use them for the precedence order; they should also be used as a way of simplifying things

Chapter 2: Naming Conventions

Naming methods, variables, and classes correctly are very important in Java and here are some of the more common conventions that, if you learn and follow, will help your code to be much clearer:

- When you are naming packages, the name should be lowercase, for example, mypackage

- All names that are representative of types have to be nouns and they are written in mixed case, beginning with an upper-case letter, for example, AudioSystem, Line

- All variable names have to be in a mixed case and start with a lower-case letter, for example, `audioSystem, line`. This makes it easier to distinguish the variables from types and also resolves any potential naming clashes.

- Where a name is representative of a constant, or a final variable, they must be in upper case and an underscore should be used separate each word, for example, `COLOR_BLUE, SAM_ITERATIONS`. Generally, use of these constants should be kept to a minimum and, more often than not, it is better to implement a value as a method, for example

```
Int     getSamIterations()     //     NOT:
SAM_ITERATIONS = 25
  {
    return 25;
  }
```

This is much easier to read and it also ensures that class values have a uniform interface

- If a name is representing a method it must be a verb and it must be mixed case with a lower-case at the beginning, for example, computeTotalLenght(), getNames(). This is exactly the same as for variables but, because they have a specific form, the Java method is already easily distinguished from the variable.

- Never use uppercase for acronyms or abbreviations when you use them as a name, for example,

Good Example:
exportHtmlSource();

Bad Example:
exportHTMLSource();

Good Example:
openDvdPlayer();

Bad Example:

```
openDVDPlayer();
```

The reason for this is because if you use all upper-case letters for a base name you will see clashes with the naming conventions we already talked about in this chapter. For example, if you had this type of variable, you would need to call it hTML or dVD and these, as you can see, are not incredibly readable. A separate problem is this - when you connect the name to another, readability is significantly compromised and the word that comes after the acronym doesn't stand out.

- All private class variables should have an underscore as a suffix, for example:

```
class Person
{
  private String name_;
  ...
}
```

Quite apart from the type and name, the most important feature of any variable is its scope. Using the underscore to indicate the scope of a class allows you to easily distinguish a class variable from a local scratch variable. The reason why this is so important is because the class variable is considered to be of higher importance than the method variable and, as such, you should treat it carefully.

One side effect of using the underscore convention is that it solves the issue of finding good variable names for the setter methods. For example:

```
void setName(String name)
{
   name_ = name;
}
```

There is an issue about whether the underscore should be a suffix or whether it should be a prefix. Both of these are used commonly but adding it as a suffix is recommended because it is the best way to preserve name readability.

Do note that the scope identification of a variable has been something of a controversial issue for some time now but, at long last, the practice of using the underscore suffix is now gaining momentum and is more accepted within the professional Java development world

- All generic variables must be named the same as their type. For example:

Good Example:
```
void getTopic(Topic topic)
```

Bad Examples:
```
void getTopic(Topic value)
void getTopic(Topic aTopic)
void getTopic(Topic t)
```

Good Example:
```
void connect(Database database)
```

Bad Examples:
```
void connect(Database db)
void connect(Database oracleDB)
```

This is good practice because it helps you to cut down on intricacy by reducing the number of names and terms used. It also makes it much easier to decide what type a variable is just by the name. If this convention doesn't fit in with your programming, it is a good indicator that you have chosen a poor type name.

- All non-generic variables are assigned a specific role and you can often name them with a combination of the type and the role. For example:

```
Point   startingPoint, centerPoint;
  Name    loginName;
```

- Use the English language for all names. This is the preferred language and the only reason you can deviate from this is if you are 100% certain that your program is never going to be seen or read by anyone who does NOT speak or read your language.

- Scope and names should match. If you have a variable that has a large scope, you can give it a longer name and, in direct contrast, variables that have a smaller scope can be given a shorter name.

If you use a scratch variable for indices or as temporary storage, do keep them short. A programmer who is reading your code and sees these variables has to be able to assume that the value of the variable not been used outside of a small number of code lines. Some of the scratch variables used for integers are `i`, `j`, `k`, `m`, `n` and commonly used for integers are `c` and `d`.

- Object names are implicit and should not be used in method names. For example:

Good Example:
```
line.getLength();
```

Bad Example:
```
line.getLineLength();
```

While the last example might look natural when it comes to class declarations, it does prove to be unnecessary, as you can see in the example.

Specific Naming Conventions

- Where an attribute is to be directly accessed, you must use the terms get and set. For example:

```
employee.getName();
employee.setName(name);

matrix.getElement(2, 4);
matrix.setElement(2, 4, value);
```

- Use the is prefix for all Boolean methods and variable. For example:

```
isSet
isVisible
isFinished
isFound
isOpen
```

Using this prefix will solve the common issue of selecting bad names for Booleans like flag or status. `isStatus` and `isFlag` just don't work and you will be forced to choose names that are more appropriate. Setter methods used for the Boolean variables should have set prefixes, as in:

```
void setFound(boolean isFound);
```

There are some alternatives that you can use instead of the prefix is and some of these work better in certain situations. There are three other prefixes you can use – has, should and can:

```
boolean hasLicense();
boolean canEvaluate();
boolean shouldAbort = false;
```

- When the name is representative of an object collection, use the plural form:

```
Collection<Point>   points;
int[]               values;
```

This immediately makes the code more readable because the user can see the variable type and the operations that are to be performed on the elements.

Chapter 3:
Commenting

Comments are incredibly useful and are used to annotate programs. These comments are there for the reader of a program, to help them to understand your program or elements of it. As a rule of thumb, the code is used to tell the computer and the computer programmer what is to be done and the comments tell the programmer why it is being done.

Comments can go anywhere in a program where whitespace can go and the Java compiler will not take any notice of them – provided they have been written correctly of course. If you don't write the comments in the correct format, not only will your program throw up errors, your code is going to look very messy and   confusing.

Let's look at the different types of comments and how they should be written:

- **Line Comments** - When you write an end-of-line comment, you should start it with a pair of forward slashes (//) and it will finish at the end of the same line the forward slashes are on. The compiler will ignore anything that comes between // and the end of the line

- **Block Comments** – Block comments start with the forward slash, then the asterisk - /* and they end with the asterisk and the forward slash - */ e.g. `/*this is a block comment*/`. Any text that appears between these two delimiters will be ignored, even if it goes over several lines.

- **Bold Comments** - These are special types of block comment that are designed with drawing attention in mind. For example:

```
/*---------------------------------------
------------------
 *   This is a block comment that will
draw attention
 *   to itself.
 *---------------------------------------
-----------------*/
```

- **Javadoc Comments** — this is another special type of block comment that starts with a forward slash, followed by a pair of asterisks - /**. These tend to be used as a way of automatically generating a class API and the following are the general guidelines for writing a Javadoc comment:

There are no actual set-in-stone rules because a good programmer will write good code that will document itself

Do be sure that your comments agree with your code; if you ever update your code, you must remember to update the comments as well.

Never write a comment that does nothing more than repeat the code. Comments should describe what the code does and why, not how it is done:

```
g++;        //  increment g by one
```

If you have code that can be confusing, either comment it or, better still, rewrite it so it doesn't look confusing.

At the start of each file, add a bold comment that contains your name, the date, what the program is for and how it should be executed:

```
/*---------------------------------------
--------------------------
 *   Author:        Sharon Stone
 *   Written:       5/12/2014
 *   Last updated:  12/17/2006
 *
 *   Compilation:   javac HelloWorld.java
 *   Execution:     java HelloWorld
 *
 *    Prints  "Hello,  World".  This  is ·the
first program
 * that everyone learns
 *   % java HelloWorld
 *   Hello, World
 *
```

```
*-------------------------------------------
-------------------------*/
```

Wherever you have a variable name, including an instance variable, that is important, comment it with the // comment and make sure that the comments are aligned vertically.

```
private double rx, ry;     //  position
private double q;          //  charge
```

Each method must be commented with a description of what the method does, including what it needs as an input, what the output will be and if there are any side effects – list them if there are. Your comment must include the parameter names.

```
/**
 *       This  method  will  use  Knuth's
shuffling algorithm to
 *      rearrange all the array elements in
[] random order
 *
 *      If  it  is  null,  it  will  throw  a
NullPointerException
 */
public static void shuffle(String[] a)
```

Chapter 4:
Java Files

Another important part of Java that you have to get right for efficiency is how you use files. For a start, every Java file must have the .java extension. In this chapter, we are also going to take a brief look at the layout of statements, classes, methods, types, variables, and numbers. Here are some more things you must get right if you want your code to be effective:

- The content of each file must be kept to no more than 80 columns.

This is the common measure for printers, terminal emulators, editors, and debuggers. If your file is shared with other developers, you must keep to this constraint otherwise, readability will be low – when you pass a file to another developer that doesn't keep to this, there will be line breaks that you possibly didn't mean to be there.

- Avoid the use of special characters like page break and TAB

Special characters cause trouble for debuggers, editors, terminal emulators and printers when they are used in files that are shared across developers and across platforms.

- If a split line is not complete, it must be made obvious. For example:

```
totalSum = a + b + c +
```

```
        d + e;

method(param1, param2,
        param3);

setText ("Long line split" +
        "into two parts.");

for (int tableNo = 0; tableNo < nTables;
     tableNo += tableStep) {
  ...
}
```

Split lines happen when you go over the 80-column constraint and, although it isn't easy to give exact rules on how to split lines, the examples above should give you some idea. As a rule of thumb:

- A break should follow a comma

- A break should follow an operator

- New lines should be aligned with the start of the expression on the line before

Statements

Import and Package Statements
- Package statements are always the first statement in a file and every file must belong to a specified package

By placing all files into a specified package, and not the default Java package, allows the object-oriented programming of Java to be enforced.

- Import statements must be placed after a package statement. They should also be sorted so that the fundamental packages come first and then group all associated packages. There should be a single blank line between each group:

```
import java.io.IOException;
import java.net.URL;

import java.rmi.RmiServer;
import java.rmi.server.Server;

import javax.swing.JPanel;
import javax.swing.event.ActionEvent;

import
org.linux.apache.server.SoapServer;
```

The location of the import statement will be enforced by Java. By sorting the packages, it makes it easier to browse through when there are a lot of imports and it also makes it easier to determine any dependencies on the present package. Grouping packages together collapses all related information into one unit and makes it less complicated. Less complicated = increased readability.

- Always explicitly list imported classes:

Good Example:

```
import java.util.List;
```

Bad Examples:

```
import java.util.*;
import java.util.ArrayList;
import java.util.HashSet;
```

When you explicitly import a class, you are providing the best documentation value for the present class and it also makes it much easier to understand the class and maintain it.

Interfaces and Classes

Make sure that all interface and class declarations are organized in the following way:

- Interface or class documentation

- Interface or class statement

- Static, or class, variables in this order – Public; Protected; Package (with no access modifiers; Private

- Instance variables in this order – Public; Protected; Package (with no access modifiers); Private

- Constructors

- Methods but there is no particular order for these

Make the location of each of the class elements predictable so it reduces complexity.

Methods
Put method modifiers in this order:

- <access> static abstract synchronized <unusual> final native

The first modifier should always be the <access modifier> if there is one:

Good Example:
```
public static double square(double a);
```

Bad Example:
```
static public double square(double a);
```

The <access> modifier is public, private or protected while the <unusual> modifier will include transient and volatile. Out of all of this, the most important thing to learn is to put the <access> modifier at the beginning – this is a very important modifier and it has to stand out in your method declarations. The order of other modifiers isn't so important.

Types

- All type conversions should be explicitly done; you should never rely on implicit conversions:

Good Example:
```
floatValue = (int) intValue;
```

Bad Example:
```
floatValue = intValue;
```

By doing this, you are indicating that you fully aware of all the different types that are in use and that the mix was deliberate.

- Attach array specifiers to type, never to the variable:

Good Example:
```
int[] a = new int[20];
```

Bad Example:
```
int a[] = new int[20]
```

Arrayness is not a feature of the variable, and it is unknown why Java allows it to be when it should be attached to the base type

Variables

- When you declare a variable, it must be initialized and you should also use the smallest possible scope to declare as well.

This is to make sure that a variable will be valid all the time. On occasion, it may be possible to initialize the variable to a proper value in the location it is declared – in this case, leave the variable uninitialized, rather than initializing it to a false value

- Never let your variables have a dual meaning

This way, all of your concepts are uniquely represented and there is a lower chance of errors popping up

- Never declare a class variable publicly

One of the main concepts of Java is to hide and encapsulate information. The declaration of a class variable as public violates that so use access functions and private variables instead. There is just one exception to this – if a class is a data structure that has no behavior. In this case, you can make the instance variable public

- Brackets beside type when you declare an array:

Good Example:
```
double[] vertex;
```

Bad Example:
```
double vertex[];
```

Good Example:
```
int[]    count;
```

Bad Example:
```
int    count[];
```

Good Examples"
```
public    static    void    main(String[]
arguments)

public double[] computeVertex()
```

There are two reasons for this. First, as we mentioned earlier, arrayness is NOT a feature of the variable, it is attached to class. Second, when an array is returned from a method, you can't have the brackets anywhere but with type.

- Keep the lifespan of a variable as short as you can

This makes it easier to control the side effects and the effects of the variable.

Loops
- Only include loop control statements in the for() construction:

Good Example:
```
sum = 0;
```

Bad Examples:
```
for (i = 0, sum = 0; i < 100; i++)
for (i = 0; i < 100; i++)
sum += value[i];
sum += value[i];
```

This increases readability and makes it easier to maintain. Always have a clear distinction of what is in the loop and what controls it

- Do not initialize a loop variable right before a loop

Good Example:
```
isDone = false;
```

Bad Examples:
```
bool isDone = false;
while (!isDone) {          //           :
    :                      //           while
(!isDone) {
}                          //           :
                           //           }
```

- Avoid using do-while loops

These are not as readable as the normal while and for loops because the conditional goes at the bottom. This means that a reader has to scan through the entire loop to understand what the scope is. You don't actually need to use a do-while loop because they can be written very easily into standard while or for loops. Cutting down on the amount of constructs you use enhances readability.

- Avoid using `continue` and `break` in loops

The only reason you should use these is they provide better readability than their more structured equivalent

Conditionals

- Avoid the use of complex conditional expressions; a better option is a temporary Boolean variable

Good Example:

```
bool isFinished = (elementNo < 0) ||
(elementNo > maxElement);
bool isRepeatedEntry = elementNo ==
lastElement;
if (isFinished || isRepeatedEntry) {
   :
}
```

Bad Example:

```
if ((elementNo < 0) || (elementNo >
maxElement)||
     elementNo == lastElement) {
   :
}
```

When you assign a Boolean variable to an expression, your program is given automatic documentation, it will be better constructed, easier to read and to debug

- Place nominal case within the `if` section and exceptions in the `else` section of the `if` statement

```
boolean isOk = readFile(fileName);
if (isOk) {
  :
}
else {
  :
}
```

Ensure that the exception doesn't cover up the normal path the execution should take; this is an important point, not just for readability but for performance as well

- Put the conditional on another line

Good Example:
```
if (isDone)
```

Bad Example:
```
if (isDone) doCleanup();
  doCleanup();
```

We do this for the purposes of debugging. If you place them on the same line, it will not be clear if the test is true or false

- Avoid using executable statements in a conditional

Good Example:

```
InputStream  stream  =  File.open(fileName,
"w");
if (stream != null) {
  :
}
```

Bad Example:

```
if (File.open(fileName, "w") != null)) {
  :
}
```

A conditional that contains an executable statement is incredibly hard to read, especially for those who are new to Java programming

Numbers

- Avoid using magic numbers in your code. Anything other than 0 and 1 is considered to be declared as a named constant

Good Example:

```
private static final int  TEAM_SIZE = 11;
:
Player[] players = new Player[TEAM_SIZE];
```

Bad Example:

```
Player[] players = new Player[11];
```

If the number on its own does not have any obvious meaning, add a named constant instead to enhance readability

- Always write floating point numbers with a decimal point and a minimum of one decimal

Good Example:
```
double total = 0.0;
```

Bad Example:
```
double total = 0;
```

Good Example:
```
double speed = 3.0e8;
```

Bad Example:
```
double speed = 3e8;

double sum;
:
sum = (a + b) * 10.0;
```

These examples show you the difference in the natures of the floating-point number and the integer. In mathematical terms, they are both different and neither is compatible with the other. Also, as you can see in the above example, this emphasizes the assigned variable type at a place in your code where it might not be obvious or evident

- Always put a digit in front of the decimal point in a floating-point constant

Good Example:
```
double total = 0.5;
```

Bad Example:
```
double total = .5;
```

The system of expressions and numbers in Java comes from the mathematical system and you should stick to the mathematical convention as far as possible for syntax purposes. Not only that, it is easier to read 0.5 than it is to read .5 and you can't muddle it up with the integer 5.

- Refer to static methods or variables using the class name, not the instance variable

Good Example:
Thread.sleep(1000);

Bad Example:
thread.sleep(1000);

These examples show that the element reference is independent of any specific instance and is static. For the very same reason, you should include the class name when you access a method or a variable from the same class.

Chapter 5:
White Space

Whitespace is another important part of coding, especially knowing how and where to use it. The general rules are:

- Always surround operators with space characters

- Follow reserved keywords with a white space

- Follow a comma with a white space

- Surround colons with white space

- Follow semi-colons that are in a `for` statement with a space character

Good Example:
```
a = (b + c) * d;
```

Bad Example:
```
a=(b+c)*d
```

Good Example:
```
while (true) {
```

Bad Example:
```
while(true){
    ...
```

Good Example:
```
doSomething(a, b, c, d);
```

Bad Example:

```
doSomething(a,b,c,d);
```

Good Example:

```
case 100 :
```

Bad Example:

```
case 100:
```

Good Example:

```
for (i = 0; i < 10; i++) {
```

Bad Example:

```
for(i=0;i<10;i++){
    ...
```

By doing all of this you are making the components of each statement stand out, increasing readability. While it isn't very easy to provide you with a full list of white space use, the above examples should give you some idea.

- Use a white space after a method name if it has another name after it

```
doSomething (currentFile);
```

This ensures that each individual name stands out and makes it easier to read. If there is no name following the method name, leave out the white space as the name will be obvious.

```
(doSomething())
```

There is an alternative method here – follow the opening parenthesis with a white space. Some who do this place another white space before the closing parenthesis but this is not really needed. It does make things stand out more though and there is no harm in doing it.

Example – two white spaces
```
doSomething( currentFile );
```

Example – one white space
```
(doSomething( currentFile);)
```

- Separate all logical units in a block with a blank line

```
// Create a new identity matrix
Matrix4x4 matrix = new Matrix4x4();

// Precompute angles for efficiency
double cosAngle = Math.cos(angle);
double sinAngle = Math.sin(angle);

// Specify    matrix    as    a    rotation
transformation
matrix.setElement(1, 1,  cosAngle);
matrix.setElement(1, 2,  sinAngle);
matrix.setElement(2, 1, -sinAngle);
matrix.setElement(2, 2,  cosAngle);

// Apply rotation
transformation.multiply(matrix);
```

By bringing in a white space between each logical unit, you are making the code more readable. Blocks are usually started with a comment, as you can see above.

- Use three blank lines to separate each method

By enlarging the space bigger than the space in the method, all of the methods will stand out

- Align variables in a declaration to the left

```
TextFile   file;
int        nPoints;
double     x, y;
```

This makes the code read better and, because they are aligned, you will better be able to see the variable types.

- Align statements wherever it makes the code more readable

```
if                  (a    ==    lowValue)
compueSomething();
else    if    (a    ==    mediumValue)
computeSomethingElse();
else    if    (a    ==    highValue)
computeSomethingElseYet();

value = (potential          * oilDensity)
/ constant1 +
        (depth              * waterDensity)
/ constant2 +
```

```
        (zCoordinateValue  *  gasDensity)
/ constant3;

minPosition         = computeDistance(min,
x, y, z);
averagePosition                          =
computeDistance(average, x, y, z);

switch (phase) {
  case  PHASE_OIL       : text  =  "Oil";
break;
  case  PHASE_WATER  : text  =  "Water";
break;
  case  PHASE_GAS    : text  =  "Gas";
break;
}
```

There are quite a few places in Java code where white space can be added to make it more readable, even if adding it causes a violation of the common guidelines Many of these places are associated with code alignment and, while it isn't easy to provide a full list of the general guidelines, the examples above should help you.

Chapter 6:
Coding Do's and
Dont's

These are just a few more do's and don'ts to help make you a more effective programmer:

- Only use a return at the end of the method, never in the middle

If you use a return in the middle, you will struggle to break the method down into smaller ones later if you want to. It will also force you to look at several exit points in one method.

- Don't use continue

Again, this makes it difficult to break a construct down into smaller ones or into methods later on, as well as forcing you to look at several endpoints in one construct

Reasoning: Using continue makes it difficult to later break the construct into smaller constructs or methods. It also forces the developer to consider more than one end point for a construct.

- Use separate lines for increments or decrements; never compound them

When you compound an increment or a decrement operator into a method call makes it difficult to read, not good for the programmer with little experience when it comes to modifying code.

Bad Example:

```
foo(x++);
```

Good Example:

```
foo(x);
    x++;
```

Bad Example:

```
y += 100 * x++;
```

Good Example:

```
y += 100 * x;
    x++;
```

- Always declare a variable as near to where it is going to be used as you can.

Example 1:

```
    int totalWide;
    int firstWide = 20;
    int secondWide = 12;
    firstWide    =      doFoo(firstWide,
secondWide);
    doBar(firstWide, secondWide);
    totalWide = firstWide + secondWide;
//  wrong!
```

Example 2:

```
  int firstWide = 20;
    int secondWide = 12;
    firstWide    =      doFoo(firstWide,
secondWide);
    doBar(firstWide, secondWide);
```

```
    int     totalWide    =      firstWide    +
secondWide;       //  right!
```

Example 3:
```
    int secondWide = 12;
    int firstWide = doFoo(20, secondWide);
    doBar(firstWide, secondWide);
    int totalWide = firstWide + secondWide;
// even better!
```

Quote:
"Any fool can write code that a computer can understand.
Good programmers write code that humans can understand."
   -   Martin Fowler

Instead of trying to document how a complex algorithm is performed, just make the algorithm more readable in the first place by adding in identifiers. This will help if the algorithm is changed in the future but the documentation is not updated.

Example:
Rather than this:
```
  if ( (hero == theTick) && ( (sidekick ==
arthur) || (sidekick == speak) ) )
```

Do this instead::
```
    boolean isTickSidekick = ( (sidekick
== arthur) || (sidekick == speak) );
    if    (    (hero    ==    theTick)    &&
isTickSidekick )
```

OR
Instead of this:

```java
    public static void happyBirthday(int age)
    {
        // If you're in the US, some birthdays are special:
        // 16 (sweet sixteen)
        // 21 (age of majority)
        // 25, 50, 75 (quarter centuries)
        // 30, 40, 50, ... etc (decades)
        if ((age == 16) || (age == 21) || ((age > 21) && (((age % 10) == 0) || ((age % 25) == 0)))))
        {
            System.out.println("Super special party, this year!");
        }
        else
        {
            System.out.println("One year older. Again.");
        }
    }
```

Do this:

```java
  public static void happyBirthday(int age)
    {
        boolean sweet_sixteen = (age == 16);
        boolean majority = (age == 21);
        boolean adult = (age > 21);
        boolean decade = (age % 10) == 0;
        boolean quarter = (age % 25) == 0;
```

```java
        if  (sweet_sixteen  ||  majority  ||
(adult && (decade || quarter)))
        {
           System.out.println("Super
special party, this year!");
        }
        else
        {
           System.out.println("One    year
older. Again.");
        }
    }
```

Chapter 7:
Common Java
Syntax Mistakes

To finish off, I am going to go over some of the more common Java syntax mistakes – getting these wrong can mess everything up!

- Keyword Capitalization

Because class names are capitalized, on occasion you might find that you do the same with a keyword, like the class `andint`. If you capitalize the first letter of this class, your compiler will not be happy and will throw up an error message – the message will depend on the keyword that you have capitalized but you will see something like this:

```
Line  nn:  class  or  interface  declaration
expected
```

- Extending strings over new lines

Sometimes your strings are going to be long but one of the most basic errors is to have a newline in the string. Again, your compiler is not going to like this and you will get an error message like:

```
Line nn: ';' expected
```

If that happens, the way around it to is to split your string into two, ensuring that neither contains a new line and then concatenate the strings with a '+':

Instead of this:

```
String s = "A rather long string which
spills over the end of the line
and will cause your compiler a big
problem";
```

with:

```
String s = "A rather long string which
spills over the end "+
"of the line and will cause your compiler
a big problem"
```

- Not putting brackets into a message with no arguments

If you have a method that has no arguments, there should be brackets after the method name. For example, if you declared the method `carryOut` without any arguments and you send a message that corresponds to the method that is to the object `objSend`, it should be written like this:

```
objSend.carryOut()
```

and not:

```
objSend.carryOut
```

- Forgetting to import packages

This is a basic mistake and one of the more common for those not experienced with Java.

When you omit the import statement from the start of the program, your compiler will give you a message like this:

```
Line nn: Class xxxx not found in type
declaration
```

That said, as java.lang is automatically imported, it doesn't require an import statement

- Muddling up static and instance methods

A very common mistake is to send a static method message to an object whereas these methods are associated with classes. For example, let's say you wanted to work out the absolute value of `intvalue` and then put it in the `int` variable; you would write:

```
 int result = Math.abs(value);
rather than:
int result = value.abs();
```

Several syntax errors can arise the most common one being:

```
Line nn: Method yyyy not found in class
xxxx.
```

In this, xxxx is the class name and yyyy is the method name

- Using the wrong case with classes

Another common error; because Java is a case sensitive language, if you use the wrong case, it won't recognize it. For example, if you write `string` and not `String`, you would get an error message like this:

```
Line nn: Class xxxx not found in type
declaration.
```

In this, xxxx is the class name with the wrong case

- Using the wrong case with variables

The same thing applies here; variables are case sensitive as well and, if you were to declare a variable `linkEdit` as an `int` and then attempted to refer to it in a class, you would get an error message like this:

```
Line nn: Undefined variable: xxxx
```

In this, xxxx is the mistyped variable name

- Using the wrong format for writing class methods

Class methods are always written in this form:
```
ClassName.MethodName(Argument(s))
```

One of the most common mistakes is to omit the name of the class and that will throw up a message like this:

```
Line nn: '}' expected
```

- Wrongly specifying a method argument

When a class is defined, each argument should be prefixed with either the name of a class

```
public void tryIt(int a, int b, URL c)
```

One of the more common errors, especially if you have come to Java from another programming language, is to not prefix ALL arguments with the type. For example, taking

```
public void tryIt(int a, b URL c)
```

This will generate an error message something like this:

```
Line nn: Identifier expected
```

- Forgetting to send a message to an object

This is another error that more common to those who are new to oop, or object oriented programming. Let's say that you have the `tryIt` method; it has `two int` arguments and it will return an `int` value. If we assume that the method is to send a message to an object, we would write it like this:

```
int   newVal   =   destination.   tryIt(arg1,
arg2)
```

The arguments, or `ints`, have been declared in this code whereas, if you write it like the next example:

```
int      newVal      =      tryIt(destination,
arg1,arg2)
```

You would get an error message like this:

```
Line nn: ')' expected
```

- Treating == as value equality

We use == to compare values with scalars and, when you apply it to an object, it will compare addresses instead. For example, let's look at an `if` statement:

```
if(newObj1 == newObj2){
...
}
```

This executes the code that is denoted by ... but ONLY when the first object has the same address as the second one. When they have different addresses but the same instance variable values, you would get an evaluation to false. This won't give you any syntax errors but you will see it when you execute the program.

- Not putting voids into methods

If a method carries out an action but doesn't return any result, you need to insert the keyword **void** before the method name. If you don't, you will likely see an error message that looks something like:

```
Line    nn:    Invalid   method   declaration;
return type required
```

- Not putting a break in a case statement

This is common in both procedural and object oriented languages. You must put a break statement at the end of a case statement if you want it to finish and then exit at the end of said case statement. If you don't put the break statement in, execution will go on to the branch beneath the one where you left out the break statement

- Not putting a return into a method

If a method is returning a value, there should be a minimum of one return statement in the method body that returns the right value type. Not doing this will result in an error message like this:

```
Line nn: Return required at end of xxxx
```

In this case, xxxx is the method that has no return

- Declaring an instance variable as private and referring to it in another class by name

When you declare an instance variable as private, you cannot access it by its name outside of the class. The only way to do this is to use a method that is declared within the class in which the instance variable resides. You will see an error message like this:

```
Line nn: Variable xx in class xxxx not
accessible from class yyyy
```

In this example, xx is a private variable, xxxx is the class where the variable is defined and yyyy is the class in which the variable is referred to

- Making use of a variable before you assign it a value

Another common error in procedural and object oriented languages. Scalers are initialized to a default value or to zero in Java so that no errors are indicated. If there are any, they are shown by an array that goes beyond its bounds, or a false result. Objects are initialized to null and if you try to reference an object that has not been initialized, it will be caught when it comes to run time.

- Assuming the incorrect value type will be generated by a message

This is another popular error when you use Java packages. One example is the use of a method that will deliver a string with digits in it but treating it as an integer. The method `getInteger` inside of `java.lang.Integer` will deliver an integer and if you try to use that value as an `int`, for example, you would see an error message like this:

```
Line     nn:     Incompatible     type     for
declaration can't convert xxxx to yyyy
```

- Mixing up prefix and postfix operators

The postfix operators, like – and ++, are used to take an old value of the variable they are applied to while the prefix operators are for the new value. So, if x is 45 and you have this statement:

```
y = ++x
```

When it is executed, x and y will both be 46. However, if this statement were to be executed:

```
y = x++
```

y would be 45 and x would be 46. You won't see these errors while you are compiling but they will become evident at run time

- Failing to remember that, if an argument is an object, it is passed to a method by reference

When you use an object as a method argument, you pass the address and not the value. This means that values can be assigned to these arguments. While treating them as a value won't particularly kick up an error, you won't be making proper use of the object oriented language

- Failing to remember that a scalar is passed to a method by value

Arguments that take the form of a scalar cannot be treated if they can be assigned to. While this won't come up as a syntax error, it will show up at run time if you write a piece of code that assumes a method has been used to pass a value to a scalar

- Not using size properly for arrays and strings

Size is an instance variable that is associated with a method when it is associated with a string and with arrays. If you were to muddle them up by writing something like:

```
arrayVariable.size()
```

or
```
stringVariable.size
```

the first example would give you an error message like:

```
Line nn: Method size() not found in class
java.lang.Object
```

And the second example would give an error message like:

```
Line  nn:  No  variable  size  defined  in
java.lang.String
```

- Using non-existent constructors

It is a common error to use an undefined constructor. For example, you could have a class in which you have the following constructors – one int, two int, three int – but you might have used a four int constructor. You would see this when compiling with an error message like:

```
Line  nn:  No  constructor  matching  xxxx
found in class yyyy
```

In this example, xxxx is the signature for the non-existent constructor and yyyy is the class where you should have defined it

- Calling constructors within a constructor that has the same name

Let's say, for example, that you defined a class called x with a `one` `int` and a `two` `int` constructor. In the `two` `int` you have referenced the argument x and this will kick up an error like:

```
Line nn: Method xxxx not found in yyyy
```

In this example, xxxx is the constructor name with the arguments and yyyy is class name where it is defined. The answer to this is to use the keyword `this`

- Assuming that a 2-dimensional array will be implemented directly in Java

This will result in wrong code, like:

```
int [,] arrayVariable = new [10,20] int
```

In java, this is illegal and you will see errors like this:
```
Line nn: Missing term
```

and:

```
Line nn: ']' expected
```

Multi-dimensional arrays can be implemented in Java but must be treated as single-dimension arrays that contain another single-dimension array, and so on.

- Treating scalars as objects

Scalars, like `float` and `int`, are not objects but you may want to treat them like objects when you deposit them into a Vector, like this:

```
Vector vec = new Vector();
vec.addElement(12);
```

This will result in a syntax error like this:

```
Line nn: No method matching xxxx found in yyyy
```

In this example, xxxx is the method name and yyyy is the class name that is expecting the object.

To solve this, convert them into objects using the object wrapper class that you will find in java.lang

- Confusing a scalar and its corresponding object type

When you use a scalar like `int` it isn't difficult to write a piece of code that assumes it can be treated like an object. For example:

```
int y = 22;
Integer x = y;
```

Will throw up an error message like:

```
Line    nn:    Incompatible    type    for
declaration. Can't convert xxxx to yyyy
```

In this example, both yyyy and xxxx are the class names

- Not typing the main method header correctly

Execution of a Java application requires a method declarations starting with:

```
public static void main (String []args){
```

If any of that line is not types properly or you omit a keyword, you will get an error at run time.

Let's say you omitted the keyword `static,` you would get an error message like:

```
Exception in thread main.....
will be generated at run time.
```

Conclusion

Thank you again for reading this book!

I hope this book was able to help you to understand how to neaten up your Java code and how to write cleaner and neater code for more effective and efficient programming.

The next step is to practice. And keep on practicing because that is the only way you will improve. The trick with computer programming languages like Java is to practice regularly because things fall out of fashion and new coding rules and styles come in every day. And if you don't keep up, you will find yourself having to go right back to basics every time.

Finally, if you enjoyed this book, then I'd like to ask you for a favor, would you be kind enough to leave a review for this book on Amazon? It'd be greatly appreciated!

Thank you and good luck!

Preview Of 'Java: Advanced Guide to Programming Code with Java'

"Java Interface"

The Java Interface is much the same as a Java class type that we talked about in the last chapter – Java Inheritance. When we talked about inheritance, we looked at a new keyword that we used in front of the word class right at the beginning of the class file. That keyword was abstract and we used it to create classes that could not be turned into objects. Now we are going to look at a brand-new type of file, the Java Interface. By the end of this chapter, you will begin to see how much structure you can give your program through the correct use of the Java interface.

What is the Java Interface?

It is important for you to understand that the Java Interface is neither object nor class. It is, is a kind of blueprint for classes but, on its own, it is not comprised of classes. This is the main difference between the Interface and an abstract class, which, as you know, is a class but you can't make it into an object. Likewise, the Java Interface also can't be turned into an object. Let's make things a little clearer with an example of what a Java Interface looks like:

```
public interface InterfaceExample {
    public void hi();
    public String getName();
    public int add (int a, int b);
}
```

Did you notice that there is no code in ANY of the methods and that each one has a semicolon at the end? The abstract is the only other method type that can end in a semicolon and not have any code and the abstract method belongs to the abstract class. In a Java Interface, you must remember that you cannot implement any of the methods and each must have a semicolon at the end of it. Why? Because it is NOT a class.

Note: The way to create an Interface is pretty much the same way as you so a class – the file name must end in .java and it must be the same name as the Interface, right down to capitalization – they must be exactly the same.

OK, we have Java file, an interface that doesn't have any working code in it so, what do we do with it? What is the point of it? To explain this better, we must go back to the abstract class. Remember, to use the abstract class we had to use the keyword extends. Look at this example:

```
public class dog extends Animal {
}
```

Did you notice that Dog extends Animal? What does that mean? It means that Dog is a subclass of the class Animal and all of its functionality comes from Animal. Now, let's make a new abstract class and let's call it FourLeggedObject:

```
public class FourLeggedObject extends Animal {
}
```

Now I want to use the Dog class to extend FourLeggedObject. Not going to happen! Why not? Because it is impossible. Why? Why can't an object extend whatever classes it wants to? For a very good reason, - Java is simply not able to handle multiple inheritances and that is exactly what this would cause.

How to Use the Java Interface

Although Java is not able to handle multiple inheritances it has kind of got around it by using Java Interface. The Interface is incredibly useful in that you can use objects to implement multiple or single interfaces. This means that your object will conform to several different Interface types. Here is an example showing the Dog class implementing multiple interfaces:

```
public   class   Dog   implements   Animal,
FourLeggedObject {
}
```

If you wanted to add one interface you would declare it but if you wanted several, you just declare them all with a comma separating each one. It really is as simple as that. However, in case you hadn't realized, the Dog class now has no choice but to implement all of the methods in Animal and in FourLeggedObject. That is the point of the Java Interface – to give you, the programmer, a "blueprint" that lets you create the object. Please note that you don't have to implement only the methods that are defined within the Interface; you can add new methods into your class in addition to the ones that you have got to have.

Note: if an error appears in your class, go back and make sure that all of the methods have been implemented in every interface that is being implemented. Also, check that the signatures are the same – return type, accessor type, capitalization of names, etc.

Java Interfaces have another benefit – the ability to put objects into collections that have different object types in them. Remember when we discussed Java Collections, a couple of chapters back? One collection that you can do this with is the ArrayList and that can be used to store specific object types. Let's look at an example where we store a group of Animals. We are going to see how Dog will implement Animal, which means that Dog is an Animal:

```
import java.util.ArrayList;
public class DogStoreExample {
    public  static  void  main  (String[]
args) {
        Dog d = new Dog();
        ArrayList<Animal>   animals   -
new ArrayList<Animal>();

    }
}
```

A collection can be told to store a group of objects based on the type of Interface, So, you don't have to store just Dogs in the ArrayList for animals you can also put them into another ArrayList – FourLeggedObject:

```
import java.util.ArrayList;
public class DogStoreExample {
    public  static  void  main  (String[]
args) {
        Dog d = new Dog();
        ArrayList<Animal>   animals   =
new ArrayLIst<Animal>();
        animals.add(d);
        ArrayList<FourLeggedObject>
        fourLeggedObjects     =     new
        ArrayList<FourLeggedObject>()
        ;
        fourLeggedObjects.add (d);
    }
}
```

What this means is that objects can be mixed and matched based entirely on the fact that they have got the same Interface. So, now we are going to store another object in the ArrayList FourLeggedObject, a chair this time (they have four legs, don't they?):

```
public         class         Chair         implements
FourLeggedObject {
}
import java.util.ArrayList;
public class DoStoreExample {
      public  static  void  main  (String[]
args) {
            Dog d = new Dog();
            ArrayList<Animal>    animals    =
new ArrayList<Animal>();
            animals.add (d);
            ArrayList<FourLeggedObject>
            fourLeggedObjects      -      new
            ArrayList<FourLeggedObject>()
            ;
            Chair c = new Chair();
            fourLeggedObjects.add ©;
      }
}
```

Now we have Chairs and Dogs in the same ArrayList because the objects we are storing are all of the type FourLeggedObject and both of the objects are FourLeggedObjects. Don't you think this is much better than having a separate ArrayList for each object type?

Have a go at using the Java Interface in the code you write. It can be of great help to start you off on creating an object when you are not quite sure how that object should be implemented – because an interface has no implementation! You can use them as building blocks if you like and I would recommend that you get to grips with these and use them to make your life easier.

In the next chapter, we are going to look at Polymorphism. Not sure what that is? Best read on then and learn how polymorphism can be used to help you create the best object-oriented Java applications.

Learn more about this book by going to Amazon.com and by typing in the words on the Amazon search box: **Java Advanced Charlie Masterson**

About the Author

Charlie Masterson is a computer programmer and instructor who have developed several applications and computer programs.

As a computer science student, he got interested in programming early but got frustrated learning the highly complex subject matter.

Charlie wanted a teaching method that he could easily learn from and develop his programming skills. He soon discovered a teaching series that made him learn faster and better.

Applying the same approach, Charlie successfully learned different programming languages and is now teaching the subject matter through writing books.

With the books that he writes on computer programming, he hopes to provide great value and help readers interested to learn computer-related topics.

www.ingramcontent.com/pod-product-compliance
Lightning Source LLC
Chambersburg PA
CBHW071141050326
40690CB00008B/1530